PLANAR LINKAGE SYNTHESIS

Ron Zimmerman

Magna Seating

Novi, MI

ISBN 979-8-218-17358-6

Printed and distributed by Ingramspark.com

TABLE OF CONTENTS

Preface 4

Planar Linkage Synthesis 5

Rigid Body Guidance 6

Point Path Generation 9

Function Generation 12

Mixed Synthesis 14

Summary 16

PREFACE

The modern scientific basis for planar linkage synthesis came from graphical methods developed in the 19th century. Graphical methods dominated until the computer revolution of the mid 20th century. Computer programs enabled the use of algebraic analytical methods. Historical graphical methods benefit from being visual and intuitive. However they are limited to fixed geometry that can be created by paper and pencil and their inherent accuracy. Once the programming is done, analytical methods are fast and accurate. However they lack the visual advantages of graphical methods. They require a graphical interface to visualize the equation solutions.

Discovery and invention are directly related to the availability of tools. Around the year 2000 computer aided design (CAD) programs introduced sketch tools that allowed the creation of constraint based moveable geometry. This means you can draw a line, point, or other geometry, and constrain it with relationships to other geometric features through formulas. These lines and points then can move as long as the defined relationships are maintained. This new tool, dynamic sketching, allows for the discovery of new linkage synthesis methods.

The new linkage synthesis method described in this book utilizes modern CAD sketching tools and allows the designer to synthesize linkages faster than ever before. The method is based on the geometric constraints imposed by the target positions on the linkage through the poles and rotation angles and is called Pole and Rotation Angle Constraints (PRC). These constraints are the necessary and sufficient conditions for achieving the target positions.

PRC combines the best of graphical and analytical methods. It has the visual intuitive advantages of graphical methods and the power and accuracy of analytical methods. PRC can be used to solve any planar linkage synthesis problem that is not overconstrained and it is the first method to completely solve the mixed synthesis problem.

CAD systems are ubiquitous in academic and industrial institutions. PRC puts into the hands of every mechanism designer a tool that will easily solve the simplest and the most difficult planar linkage synthesis tasks. PRC allows the designer to see the entire family of solutions by simply dragging a point around on a plane. The designer can then choose the best solution in a matter of seconds. PRC is so easy to learn and apply that novice engineers will quickly be able to learn and use it.

This book specifically covers dimensional synthesis. It can be used as a stand alone text, or as a substitute for the chapter on dimensional synthesis in a more comprehensive book.

PLANAR LINKAGE SYNTHESIS

Kinematics is the study of motion. In this book, we will learn to use our knowledge of kinematics to devise rules for creating a planar linkage mechanism with the motion we desire.

Design Tasks

The purpose of a mechanism is to perform a useful task. The process of designing the best possible mechanism for a given task involves three different research activities: type synthesis, dimensional synthesis and analysis. [1]

The goal of type synthesis is to determine what type of mechanism is best suited to perform the task. Considering things like the design task, cost, ease of use, and reliability, what type of mechanism is best suited to accomplish the design task? How many links and pivots should there be? And what configuration should they be in?

Once a mechanism type is selected, dimensional synthesis can begin. Dimensional synthesis is determining the exact length of each link and location of each pivot required to perform the task. For most design tasks, there is a large family of solutions. The designer's task is to look at the family members and determine which performs the task best. Determining which is best gets into the third phase which is analysis. This involves looking at things like package space, forces, transmission angles, velocity, acceleration and energy.

Designing linkages is a highly non-linear process. This means it can be non-intuitive. Due to this, many iterations of these three steps is frequently required before a final design is selected. You may find in the analysis phase that the forces are too high requiring you to go back and reconsider other linkage dimensions or another type of mechanism.

Design Problem Classification

There are three different classes of design tasks that linkage mechanisms can be designed to do. They are called *rigid body guidance*, *point path generation* and *function generation*. [2]

The task of rigid body guidance, also called motion generation, is to move a rigid body through a series of desired positions. For example, the trunk lid of many sedans is the coupler link of a fourbar linkage. The links control the motion of the lid when opening and closing the trunk. See Figure 1.

The point path task is to move a point on a coupler link through a series of points without regard to the orientation of the link. For example an elliptical exercise machine is designed to guide the users foot along prescribed natural path. See Figure 2.

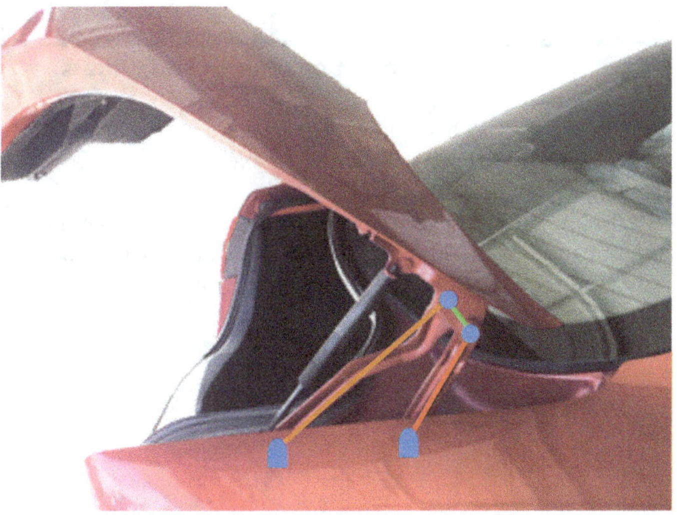

Figure 1   Rigid Body Guidance, Trunk Lid

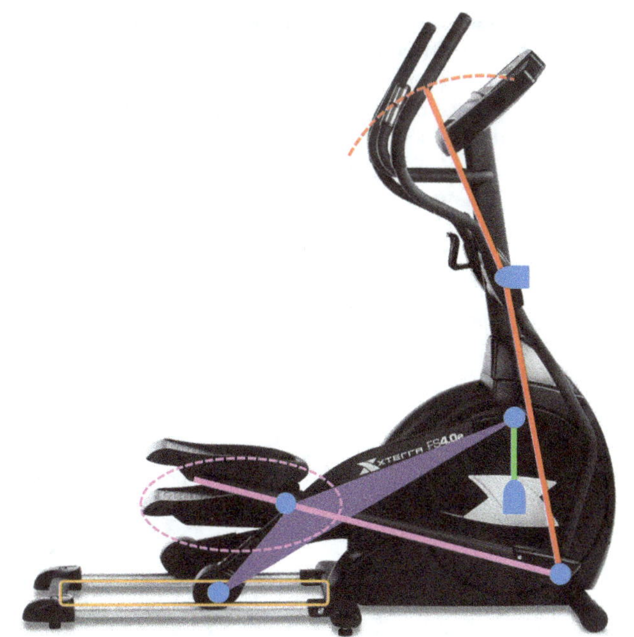

Figure 2   Point Path Guidance, Elliptical Exercise Machine

The function generation task is to control the angular relationship between guiding links. For example, the windshield wipers of a car are guiding links connected by a coupler so that they move in a synchronized way to clean your windshield. See Figure 3.

Figure 3   Function Generation, Windshield Wipers

The terms rigid body guidance and point path generation are self-explanatory. The term function generator comes from the fact that before the advent of electronic computers and calculators, linkage mechanisms were used to estimate mathematical functions in which one link pointed to a scale for the function input and the other pointed to a scale for the function output. See Figure 4. [3]

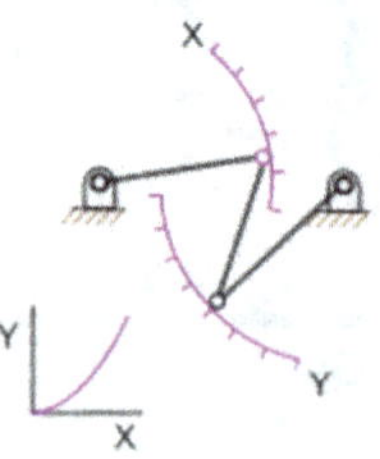

Figure 4   Function Generator

A fourth class of problems is called *mixed synthesis*. As the name implies it is a mechanism design problem in which some combination of *rigid body guidance, point path* and *function gen-*

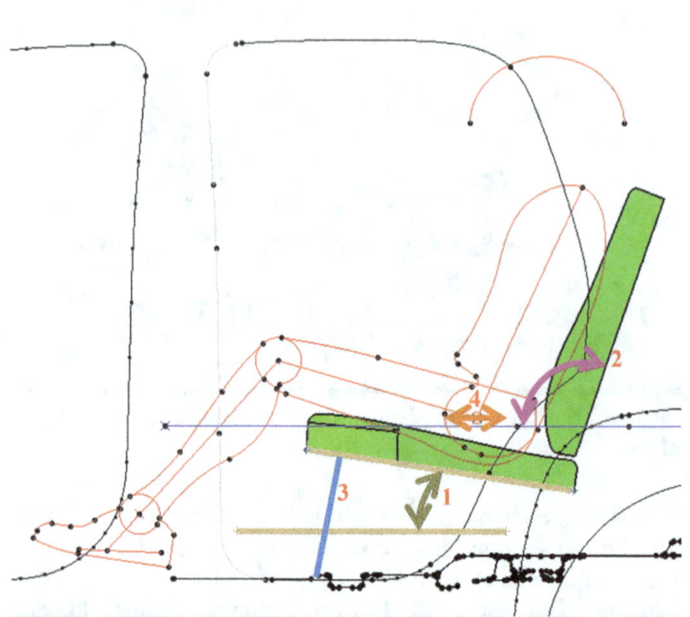

Figure 5   Mixed Synthesis, Automotive Seat

eration is required. For example, the mechanism for the second row folding seat of a utility vehicle is a fourbar linkage that has several design requirements. (The seat back and front leg are the input links and the seat cushion is the coupler.) See Figure 5.

1   Control the lower seat cushion position in the seating position and stowed position. (rigid body guidance)

2   Control the seat back angle in the design and stowed position. (function generation)

3   Control the front leg angle in the seating position in order to manage the crash loads. (function generation)

4   Control the motion of the occupant's hips when reclining so that they do not move vertically. (Point path generation.)

Figure 6 contains linkage terminology that will be used in the upcoming discussion. It shows a fourbar linkage in two positions.

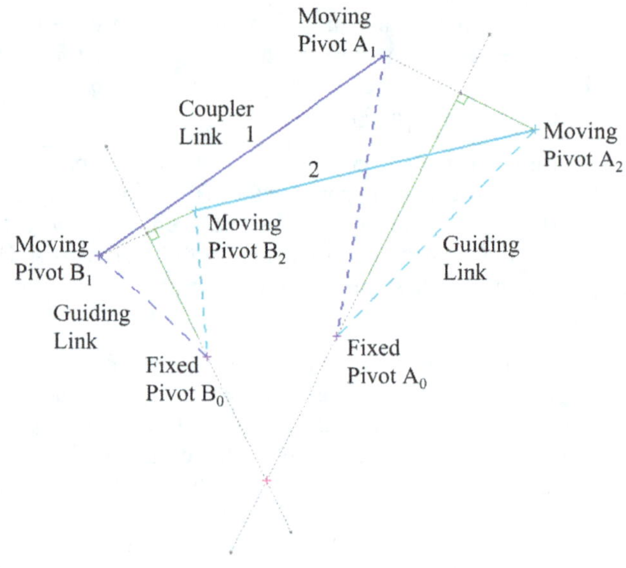

Figure 6   Linkage Terminology

RIGID BODY GUIDANCE [4]

Rigid Body Guidance: Two Positions

We will first investigate the solutions to the rigid body guidance problem. Consider a rigid body in two target positions as shown in Figure 7. The simplest mechanism would be a link with a single pivot that could carry our rigid body from position 1 to position 2. Such a pivot would have to be equidistant from A_1 and A_2, B_1 and B_2 and every other pair of corresponding points in positions 1 and 2. The perpendicular bisector of points A_1 and A_2 contains all the points equidistant from A_1 and A_2. Similarly we can find the points equidistant from B_1 and B_2 by drawing their perpendicular bisector. As long as the two perpendicular bisectors are not identical or parallel, they will intersect at a point. This point is called the **Pole**. [5] It is designated P_{12} indicating it is the pole for positions 1 and 2. Similarly the rotation angle is called Φ_{12}. The order of the subscripts indicates that the rotation is from position 1 to position 2.

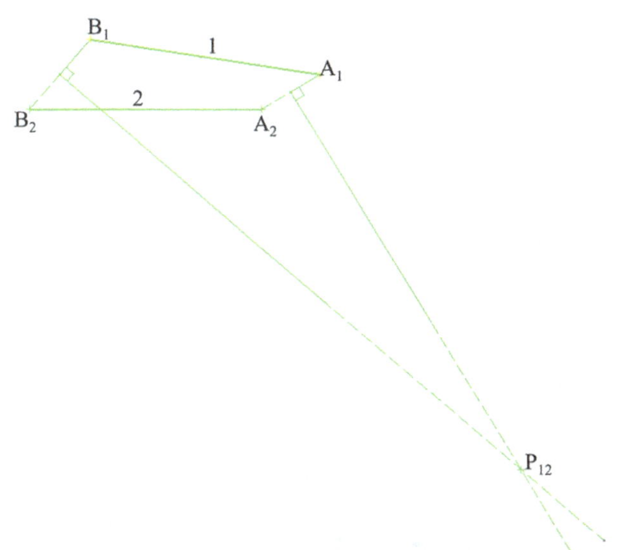

Figure 7   Two Positions of a Rigid Body

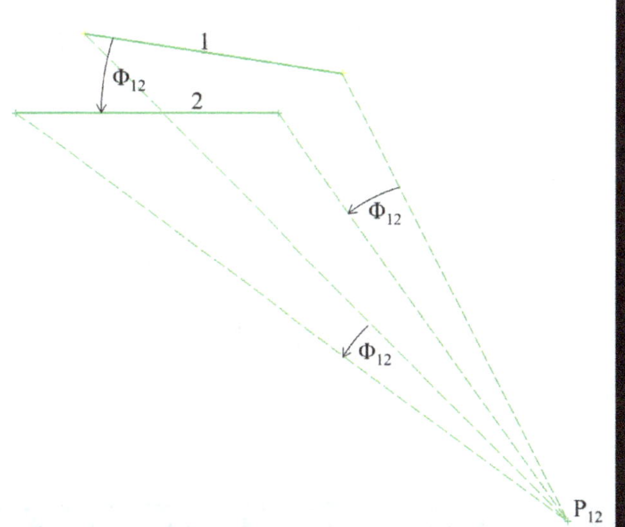

Figure 8 Pole and Rotation Angle

Two special properties can be observed by investigating the pole.

1   Target positions 1 and 2 can be reached by a pure rotation around the pole.

2   The rotation angle can be measured in several places shown in Figure 8.

This leads to the following definitions.

Pole: *For any two positions, it is the point around which the rigid body can rotate and exactly hit both positions.*

Rotation angle: *The angle that the rigid body rotates between the two specified positions. The rotation is from the position of the first suffix to the position of the second suffix.*

It is possible that this simple pivoting link solution will not be suitable for our application. Perhaps the pole is outside of the allowable space. Then we can consider creating a fourbar linkage whose coupler will carry the rigid body between the two target positions.

If point A is selected as a moving pivot (A_1 in position 1 and A_2 in position 2), then the fixed pivot A_0 must be on the perpendicular bisector of A_1-A_2 because A_0-A_1 and A_0-A_2 must be the same length. They represent the length of a guiding link. Note the fixed pivot A_0 can be anywhere on the perpendicular bisector. The perpendicular bisector also passes through the pole. If lines from the pole are drawn through points A_1 and A_2 these lines are separated by the rotation angle and are bisected by the line from the pole through A_0. These are the properties of the pole noted earlier. These relationships also exist for point B and every other point on the coupler link which leads to the following theorem.

Given two positions of a coupler link, moving and fixed pivots are on lines through the pole separated by half the rotation angle. The rotation from the moving pivot line to the fixed pivot line is in the same direction as the coupler rotation angle.

Lines from the pole through a moving pivot are called moving pivot lines and lines from the pole through a fixed pivot are called fixed pivot lines. The angle from the moving pivot line in position 1 to the fixed pivot line is half the rotation angle Φ_{12}. The order of the subscripts determines the direction of rotation; in this case, from position 1 to position 2. Φ_{21} is the rotation angle from position 2 to position 1. Note that the fixed and moving pivot lines pass through the pole and pivots can be selected from points on the line on either side of the pole. See Figure 9.

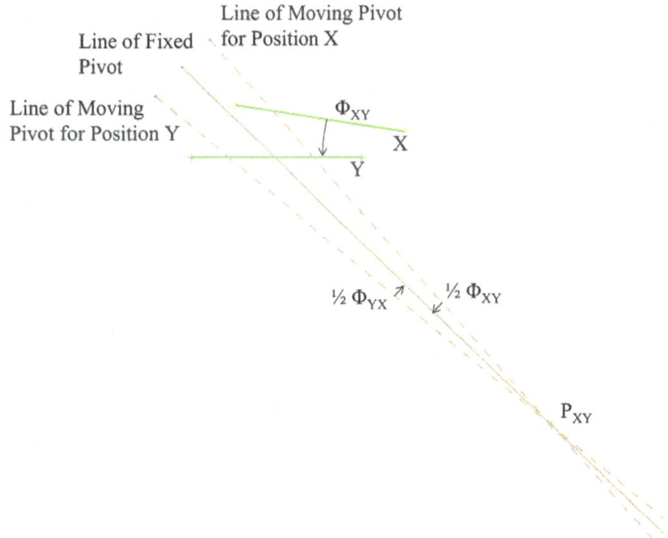

Figure 9   Fixed and Moving Pivot Lines

The method for identifying guiding links for the 2 position rigid body guidance problem can be summarized in these steps as shown in Figure 10.

1   Identify and label the two target positions.

2   Locate the pole for those two positions.

3   Draw two lines through the pole separated by half the coupler rotation angle.

4   Rotate this pair of lines to any desired angle.

5   Select any point on the moving pivot line to be the moving pivot.

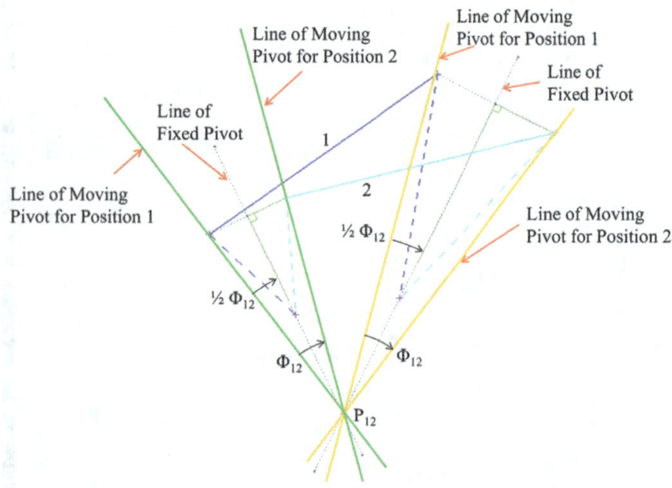

Figure 10   Linkage Found From Fixed and Moving Pivot Lines

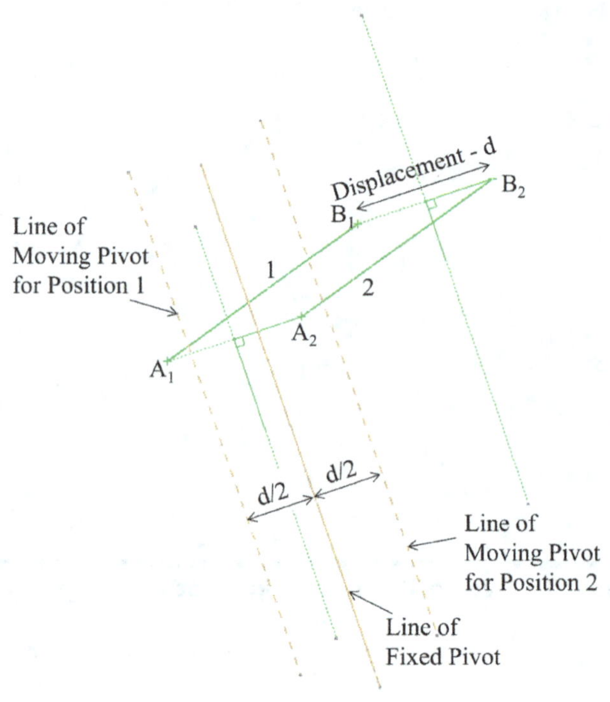

Figure 11   Fixed and Moving Pivot Lines for Parallel Positions

6   Select any point on the fixed pivot line to be the fixed pivot.

This completes one guiding link. Repeat steps 3-6 for the second guiding link. Animate the mechanism using the selected guiding links to verify the motion is suitable.

Each guiding link is defined by four parameters which are the coordinates of its end points. Given two target positions of the coupler, among the four parameters, there are three free choices for each guiding link given in steps 4-6.

There is a special case in which the rotation angle between the two target positions is zero. When the perpendicular bisectors are drawn they are parallel lines. This means:

1   The pole is infinitely far away in the direction of the perpendicular bisectors.

2   The rotation angle is zero.

3   The two positions can be reached by sliding along a line parallel to $A_1 A_2$.

Our theorem becomes: Moving and fixed pivots are on lines parallel to the perpendicular bisector and are separated by half the distance between the two positions. The direction from the moving pivot line to the fixed pivot line is in the same direction as the motion between the two positions. See Figure 11.

In order to simplify the language, the remaining discussion will assume the rotation angles are not zero. The case where they are zero is not precluded, it simply requires the adaptation just mentioned.

The moving and fixed pivot lines when rotated around the pole are the set of all the solutions to the two positions rigid body guidance problem.

When synthesizing a mechanism, it is best to think of the coupler link as a plane and the lines $A_1 B_1$ and $A_2 B_2$ simply as a way to indicate the position of the plane and unrelated to the moving pivot locations that will be selected.

Rigid body guidance: Three or more positions

Suppose we want our coupler link to pass through three positions labeled 1, 2 and 3. There are three poles (P_{12}, P_{23}, and P_{13}) and three rotation angles (Φ_{12}, Φ_{23}, and Φ_{13}). Our task is to find all the fixed and moving pivot solutions that satisfy these three positions and decide which one is the best for our application. P_{12} and the fixed and moving pivot lines through it represent all the possible fixed and moving pivot solutions for positions 1 and 2. The same applies to P_{23} and P_{13}. In general we can say P_{xy} and the fixed and moving pivot lines through it represent all the possible fixed and moving pivot solutions for positions x and y. When we find moving pivot locations they will be associated with one position of the coupler link, so we must decide which position of the moving body we want to do the synthesis on. If we select position 1 to do the synthesis on then we will use the poles with 1 in their subscript and the moving pivot lines associated with position 1.

Since P_{12} and its moving and fixed pivot lines represent all the solutions for positions 1 and 2 and P_{13} and its moving and fixed pivot lines represent all the solutions for positions 1 and 3 then their intersection represents all the solutions for positions 1, 2 and 3. The two fixed pivot lines intersect at possible fixed pivot locations and the two moving pivot lines for position 1 intersect at corresponding moving pivot locations. A line drawn between the two intersection points represents all the possible guiding links as they are moved in the plane. Either the fixed or moving pivot can be moved in the plane to any location. See Figure 12.

For the three position synthesis problem, there are two free choices available from among the four parameters defining each guiding link. Each additional target position imposes a constraint

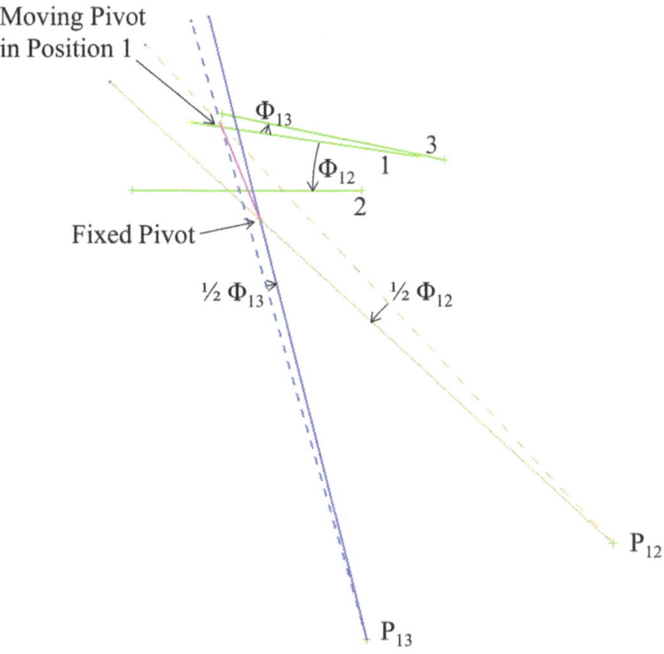

Figure 12   Solution for Three Positions

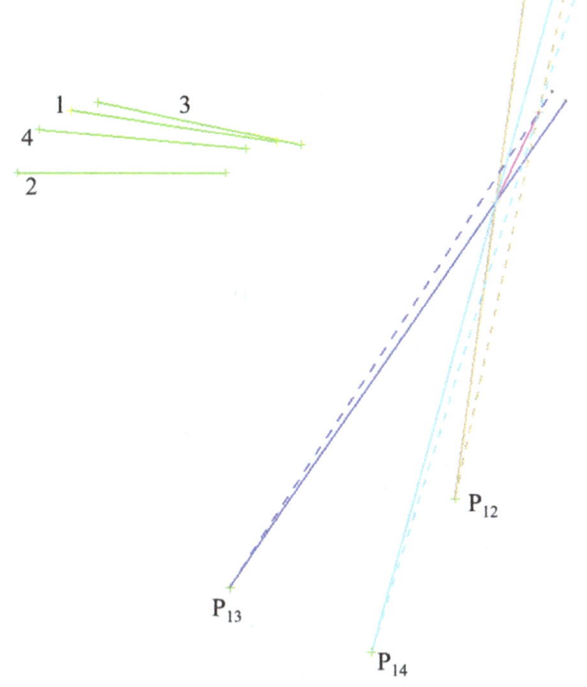

Figure 13   Solution for Four Positions

or limitation on the solution set. These constraints are manifest through the poles and rotation angles. Synthesis by this method is called Pole and Rotation angle Constraints (PRC).

For the four position rigid body guidance problem, P_{14} and its moving and fixed pivot lines are added. The moving pivot line for position 1 is constrained to intersect the moving pivot lines from P_{12} and P_{13} at a common point. The fixed pivot lines from the three poles are also constrained to intersect at a common point. See Figure 13. There are an infinite number of solutions to this problem but, this solution set is one dimensional so there is only one free choice available when defining each guiding link. As the fixed or moving pivot is moved in the plane, the circle and center point curves first identified by Ludwig Burmester are traced. [6]

For the five position rigid body guidance problem there are 0, 2 or 4 possible guiding links. These solutions are found by identifying P_{15} and drawing its moving and fixed pivot lines. The moving pivot line for position 1 is constrained to intersect the three other moving pivot lines. When this is done, the fixed pivot line from P_{15} will not likely pass through the intersection of the other fixed pivot lines. The moving pivot must be moved until locations are found where the four fixed pivot lines also intersect at a common point. These are the possible guiding links for the 5-position rigid body guidance problem. Five positions of the coupler link is the most for which a solution can be expected. The number of free choices available for each number of target positions is given in Table 1.

Table 1   Number of Free Choices Available

Number of Positions	Free Choices
2	3
3	2
4	1
5	0

Defects

Linkage synthesis by this method guarantees that the linkage can be assembled in each target position. However, it does not control what happens in between those positions. Once both guiding links are selected, it is required to animate the mechanism to verify that the motion of the coupler between the targeted positions is as intended. There are two types of defects that may occur in fourbar linkages. The first is called a branching or circuit defect. A fourbar linkage of given link lengths can generally be assembled in two ways. These are called branches or circuits. See Figure 14. If all of the target positions are not on the same circuit then the linkage cannot reach all the positions without disassembly and reassembly. In this case, new guiding links must be selected. The other type of defect is called an order defect and can occur when 4 or more target positions are required. As the name implies, this type of linkage will pass through the targeted positions, but not in the desired order.

After a suitable kinematic solution is obtained, it may be necessary to do additional iterations when other factors are considered such as forces, transmission angles and acceleration.

POINT PATH GENERATION [7]

Point Path Generation: Two positions

We will now consider the point path generation problem using the knowledge we have of pole and rotation angle constraints. The point path task is to move a point on a coupler link through a series of points without regard to the orientation of the coupler. First, we will consider two target points and ask the following questions:

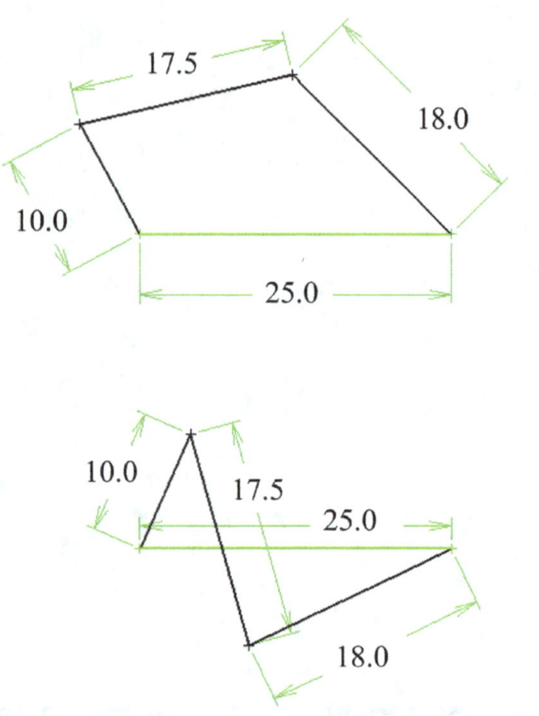

Figure 14   Open and Crossed Branches

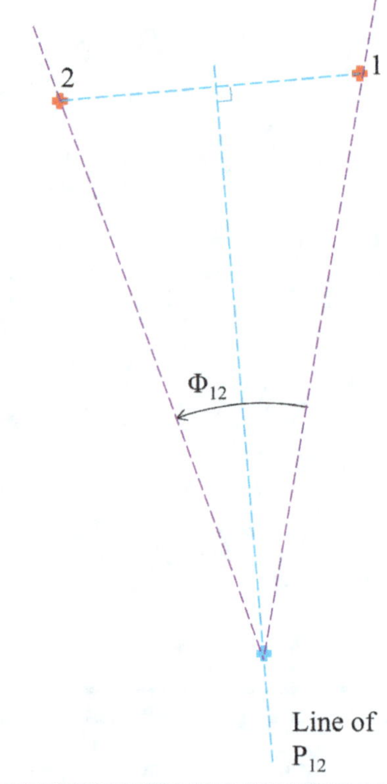

Figure 15   Pole and Rotation Angle for Point Path Problem

1   Where is the pole?

2   What is the rotation angle?

We will call the points 1 and 2. The pole (P_{12}) must be on the perpendicular bisector of the two points. In this case the pole is not fixed but can be anywhere on the line. During the synthesis process we will allow the pole to float on the line. If lines are drawn from the pole to the target points the angle between them is the rotation angle (Φ_{12}). The rotation angle varies with the location of the pole. We will simply monitor and measure the rotation angle. See Figure 15.

For the point path problem, both guiding links must be designed at the same time. We will draw two pairs of moving and fixed pivot lines from the pole each separated by half the rotation angle, one pair for each guiding link. Points representing the moving and fixed pivots are added, one to each line. These points are free to move anywhere on their line. The two fixed pivots and two moving pivots can be moved to suitable locations based on the design requirements. As the pole moves along the perpendicular bisector the rotation angle will update automatically. These 4 fixed and moving pivot lines represent all the solutions to the 2 position point path problem. See Figure 16.

Eight variables define the endpoints of the two guiding links. For the two position point path problem we have 7 free choices. They are:

1   Location of the pole on the perpendicular bisector.

2 & 3   Orientation of each set of fixed and moving pivot lines.

4 - 7   Location of each fixed and moving pivot on its line.

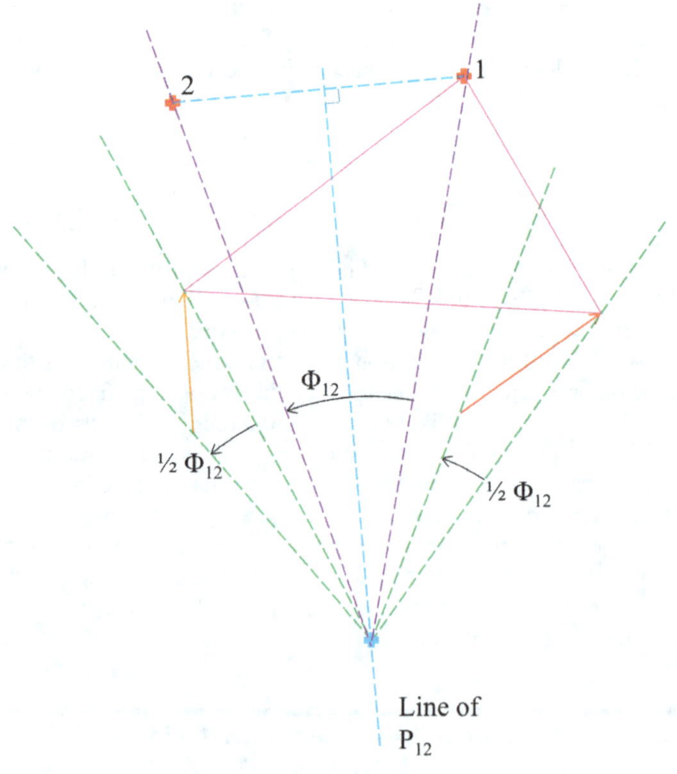

Figure 16   Fixed and Moving Pivot Lines

For the point path problem: we have the following theorem.

Given a point on a coupler link in two different positions, moving and fixed pivots are on lines through the pole separated by half the rotation angle. The rotation from the moving pivot line to the fixed pivot line is in the same direction as the coupler rotation angle.

The method for identifying guiding links for the 2 position point path problem can be summarized in these steps.

1   Identify and label the two target positions of the point.

2   Locate the pole for those two points.

3   Draw two lines through the pole separated by half the coupler rotation angle for each guiding link.

4   Rotate each pair of lines to any desired angle.

5   Select any point on each moving pivot line to be a moving pivot.

6   Select any point on each fixed pivot line to be a fixed pivot.

Animate the mechanism using the selected guiding links to verify the motion is suitable.

Point Path Generation: Three or More Positions

For three target points (1, 2 and 3) we will add the pole P_{13}, measure the rotation angle(Φ_{13}), and draw 2 pairs of fixed and moving pivot lines separated by half the rotation angle from it. The four fixed and moving pivot lines from P_{13} represent all the solutions for points 1 and 3. The solutions for points 1, 2 and 3 are where the fixed and moving pivot lines from P_{12} intersect with the fixed and moving pivot lines from P_{13}. See Figure 17.

If additional points are required to define the desired path, the same procedure of adding points with their poles and fixed and moving pivot lines is followed. The solution pivots for all the target points are where the fixed pivot lines intersect and where the moving pivot lines intersect. The poles must have a common subscript. The common subscript is the position of the coupler on which you are finding the moving pivots. See Figure 18. This can be continued for up to 9 target points which is the limit for exact synthesis. In the limit, poles P_{12}, P_{13}, P_{14}, P_{15}, P_{16}, P_{17}, P_{18}, and P_{19} would be used with rotation angles Φ_{12}, Φ_{13}, Φ_{14}, Φ_{15}, Φ_{16}, Φ_{17}, Φ_{18}, and Φ_{19} to solve a 9 position point path synthesis problem solving for the moving pivots in position 1. The chart in Table 2 shows the design freedom available depending on how many target points there are.

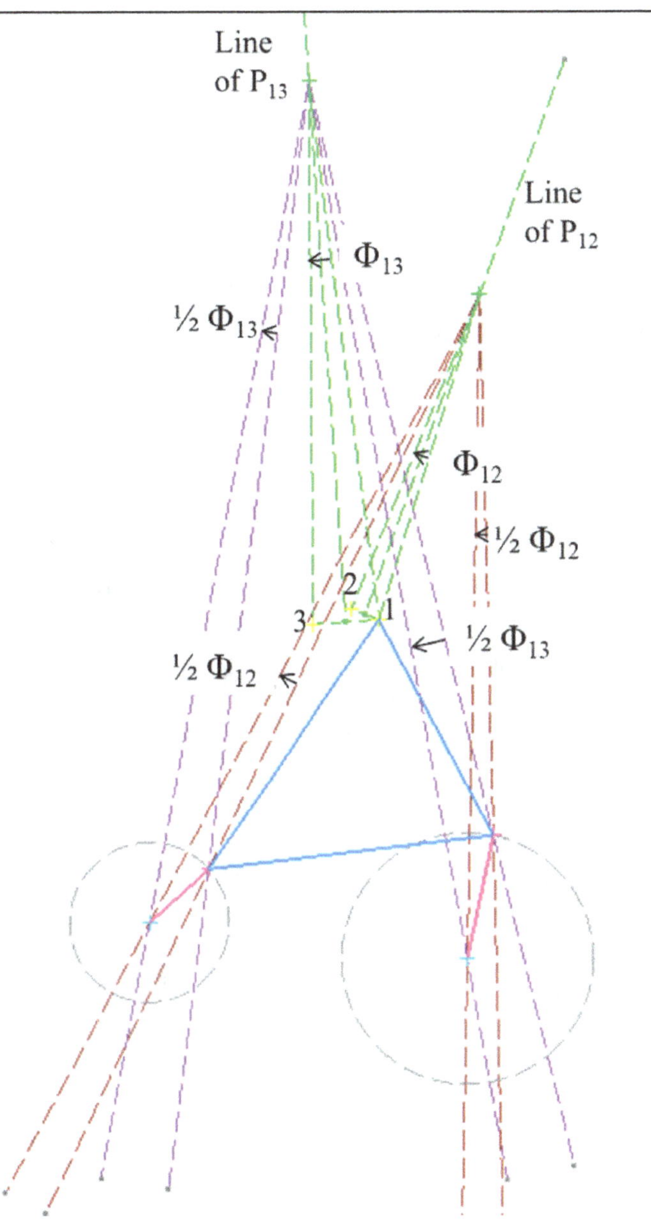

Figure 17   Three Point Solution

Table 2   Number of Free Choices Available

Number of Target Points	Free Choices
2	7
3	6
4	5
5	4
6	3
7	2
8	1
9	0

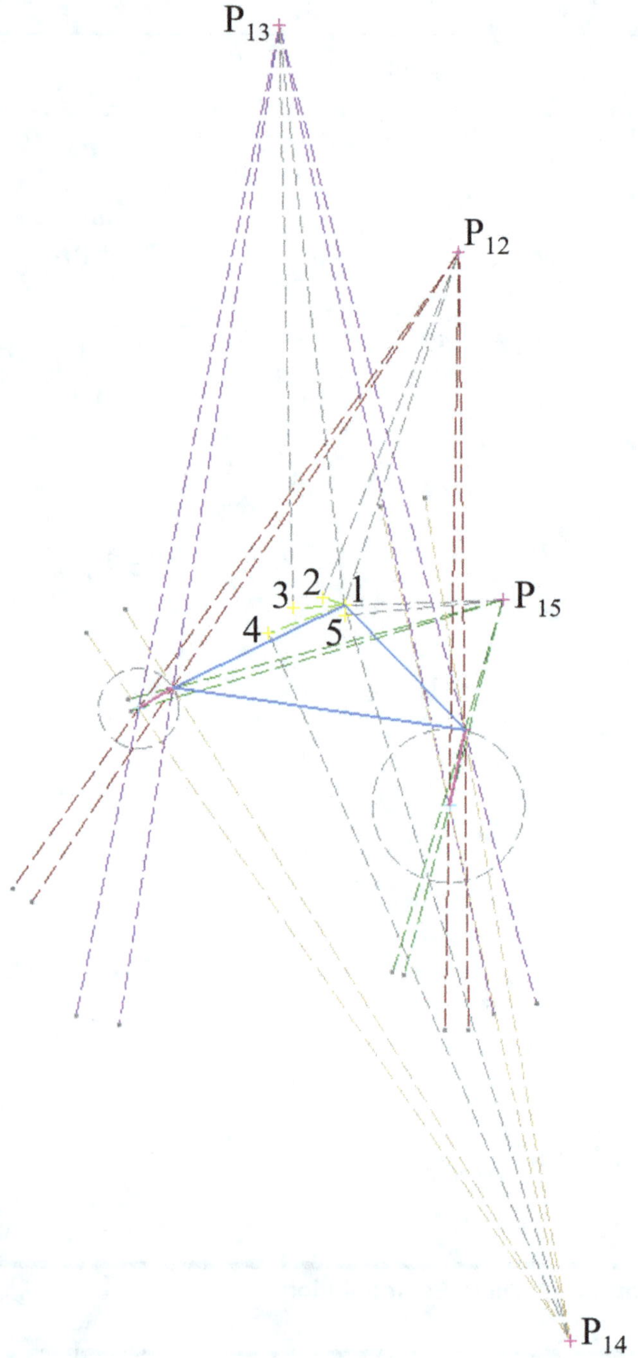

Figure 18   Five Point Solution

FUNCTION GENERATION [8]

Function Generation: Two positions

The function generation task is to control the angular change in the guiding links A and B. For two positions this means we want link A to rotate θ degrees while link B rotates ψ degrees. We will call the positions of the guiding links 1, 2, 3, etc. First, we will consider two target positions or one angle change for each link and ask the following questions:

1   Where is the pole?

2   What is the rotation angle?

The pole (P_{12}) must be on the perpendicular bisectors of θ_{12} and ψ_{12}. These are the fixed pivot lines. Note that the locations of the fixed pivots and the pole can be moved anywhere in the plane. In addition, the coupler rotation angle Φ_{12} must be the same for both links. So we will draw moving pivot lines from the pole separated by half the rotation angle. We don't care what the coupler rotation angle is so it is allowed to vary. Once these constraints are made, the fixed and moving pivots can be moved to any suitable location in the plane. See Figure 19. This construction gives all the solutions to the 2 position function generation problem. For the two position problem there are 7 free choices among the 8 variables which define the fixed and moving pivots. They are:

1 & 2   Location of the pole in the plane

3 - 6   Location of the two fixed pivots

7   The rotation angle or the location of one moving pivot.

For the function generation problem we have the following theorem.

Given an angular change for two guiding links, moving and fixed pivots are on lines through the pole separated by half the rotation angle. The rotation from the moving pivot line to the fixed pivot line is in the same direction as the coupler rotation angle.

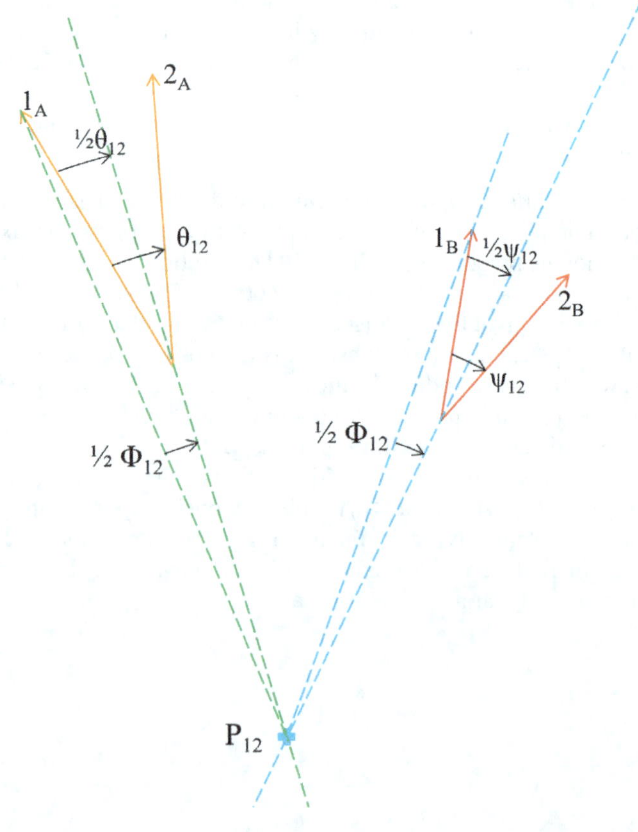

Figure 19   Function Generation—Two Positions

The method for identifying guiding links for the 2 position function generation problem can be summarized in these steps.

1 Identify and label each guiding link and its target angle change.

2 Locate the pole for those two angle changes.

3 Draw a moving pivot line for each link through the pole separated by half the coupler rotation angle.

4 Choose the two fixed pivot locations anywhere in the plane.

5 Select the coupler rotation angle and locate the moving pivots on each link.

Animate the mechanism to verify the motion is suitable.

Function Generation: Three or more positions

The three position function generation problem adds another angle change for each link. We will call these θ_{13} and ψ_{13}. We will locate P_{13} at the intersection of the bisectors of θ_{13} and ψ_{13}. The coupler rotation angle is Φ_{13}. We now draw moving pivot lines separated by half the rotation angle ($\frac{1}{2}\Phi_{13}$) from the fixed pivot lines. Where these moving pivot lines intersect the moving pivot lines for positions 1 and 2 we find a solution for all three positions. See Figure 20. The fixed and moving pivots can be moved to suitable locations. For each additional position, another pole and set of fixed and moving pivot lines are added and solutions for all the positions are where all the moving pivot lines intersect. See Figure 21.

For the pure function generator problem, the maximum number of positions for which a solution can be expected is 5. There is also an infinite number of solutions for the 5 position

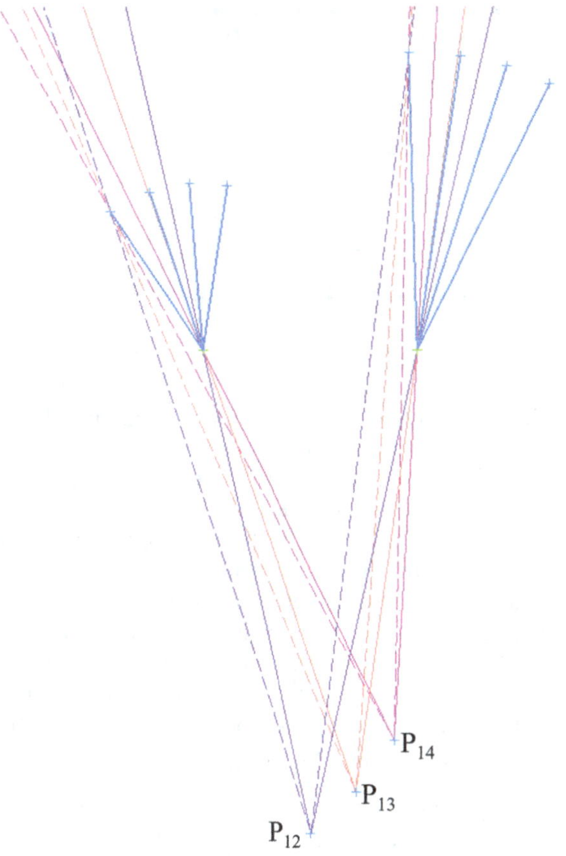

Figure 21   Function Generation—Four Positions

problem. This is because there are four attributes of the pure function generator problem that result in redundant solutions. They are:

1 Scale. This is the size of the mechanism. If all the link lengths increase or decrease by some percentage, it does not change the angular relationships between the input links during motion.

2 Orientation. If the entire mechanism is rotated in the plane, it does not change the angular relationships between the input links during motion.

3 & 4   Reference angle of guiding links. Where the angle change is measured on the link is irrelevant

These four choices are arbitrary and do not affect the solution. This leads to the following summary chart (see Table 3) indicating the number of free choices for each number of target positions.

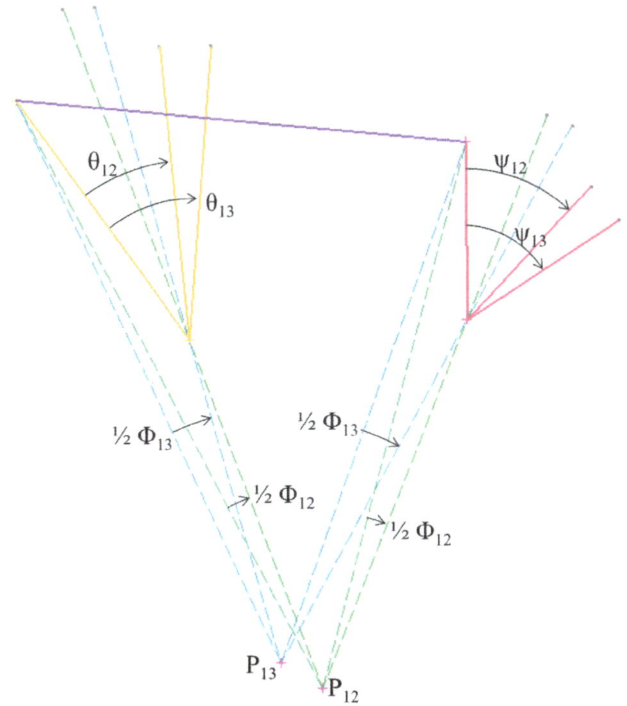

Figure 20   Function Generation—Three Positions

Table 3   Number of Free Choices Available

Positions	Arbitrary choices	Free choices
2	4	3
3	4	2
4	4	1
5	4	0

MIXED SYNTHESIS [9]

The final class of problems is a mechanism design problem in which some combination of rigid body guidance, point path and function generation is required. We will use the following notation to classify these problems: M-P-F. Where M is the number of target rigid body guidance (motion generation) positions for the coupler, P is the number of target path points on the coupler and F is the number of function generation positions. The number of body guidance and path generation positions are counted exactly as explained above. Here, the function generation positions are counted once for each target angle change of each guiding link. Specifying a travel angle for both the input and output links counts as two constraints i.e. F = 2.

If we have a synthesis problem with two coupler positions, one point path position and one guiding link angle change the problem will be classified as 2-1-1 as shown in Figure 22.

There are multiple possible configurations of targets within most combinations. For example within the 3-2-2 combination the three coupler poses and two points can be in any order and the two input link angle requirements can both be on one link, or one on each link and can correspond to any of the coupler poses or

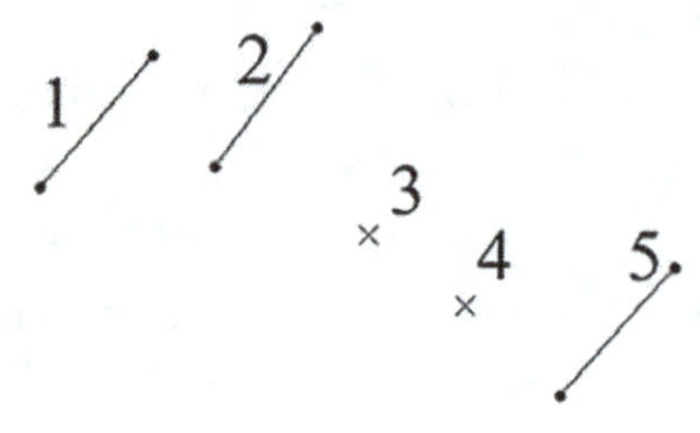

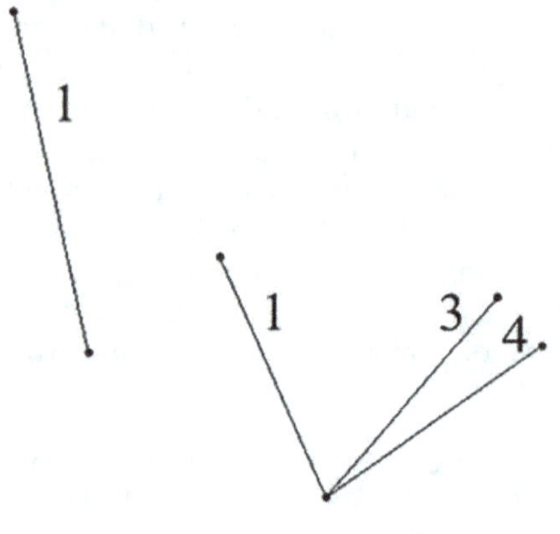

a) **Configuration One**

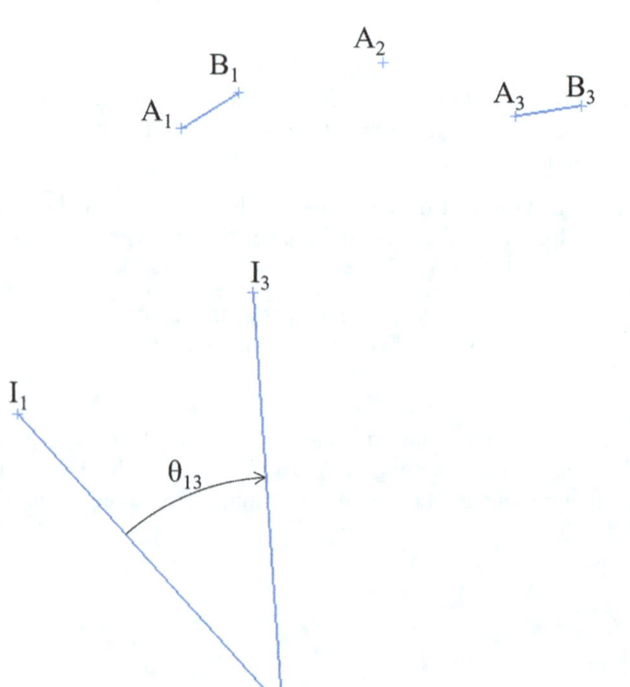

Figure 22   Mixed Synthesis Example 2-1-1

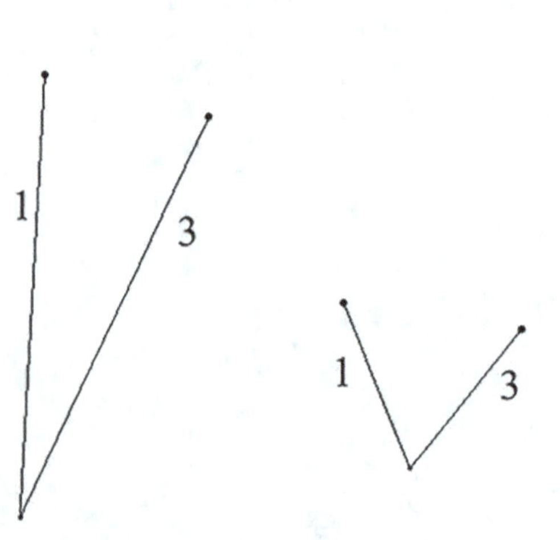

a) **Configuration Two**

Figure 23   Mixed Synthesis Two 3-2-2 Configurations

points or be independent of the coupler position. Two different 3-2-2 configurations (out of over a hundred possible permutations) are shown in Figure 23.

Solving these types of problems simply requires that we apply the principles we already know about the body guidance, path and function generation problems. Apply the pole and rotation angle constraints by identifying the poles and rotation angles and then drawing fixed and moving pivot lines.

The solution steps for the 2-1-1 problem of Figure 22 is given here:

1 Apply constraints for the rigid body guidance problem of positions 1 and 3. Find pole P_{13} and rotation angle Φ_{13}. Draw 2 sets of fixed and moving pivot lines from the pole; one for the input link and one for the output link.

2 Apply the function generator requirement that the input link rotates through the angle θ_{13} while the coupler moves from positions 1 to 3. The angle between the input link in position one and the fixed pivot line from P_{13} is ½ θ_{13}.

3 Apply the point path constraint for coupler point 2. Find the line of P_{12}. Measure the rotation angle Φ_{12} for a given P_{12} location. Note that Φ_{12} is a variable. It varies with the distance of P_{12} from the target points. Draw 2 sets of fixed and moving pivot lines from the P_{12}, one for the input link and one for the output link.

4 Constrain the fixed and moving pivots of the input and output links to be at the intersection of their respective fixed and moving pivot lines. Steps 1-3 can be done in any order. See solution in Figure 24.

Since this problem is underconstrained, the guiding links can be dragged around in the plane until optimal pivot locations are found. This procedure of successively adding body guidance, path and function generator constraints can be applied to any mixed synthesis problem that is not overconstrained.

Steps to solve a 3-2-2 problem as in Figure 23 - configuration 1.

1 Locate the poles P_{12}, P_{13}, P_{14} and P_{15}.

2 Draw two sets of fixed and moving pivot lines from each pole separated by half their rotation angle.

3 For each guiding link, constrain the fixed and moving pivot lines to intersect at one point. See solution in Figure 25.

The 3-2-2 problem is at the design limit meaning there are a finite number of solutions. For these types of problems, the last constraint may be difficult to make depending on the capabilities of your CAD system and the smoothness of the target positions.

For the mixed synthesis case the design limit is defined by this equation.

$$10 = 2M + P + F$$

In this equation, M cannot be zero. The first target coupler position imposes no restriction on the pivot locations. If no coupler body poses are specified, then one can be added without affecting the solution. [10]

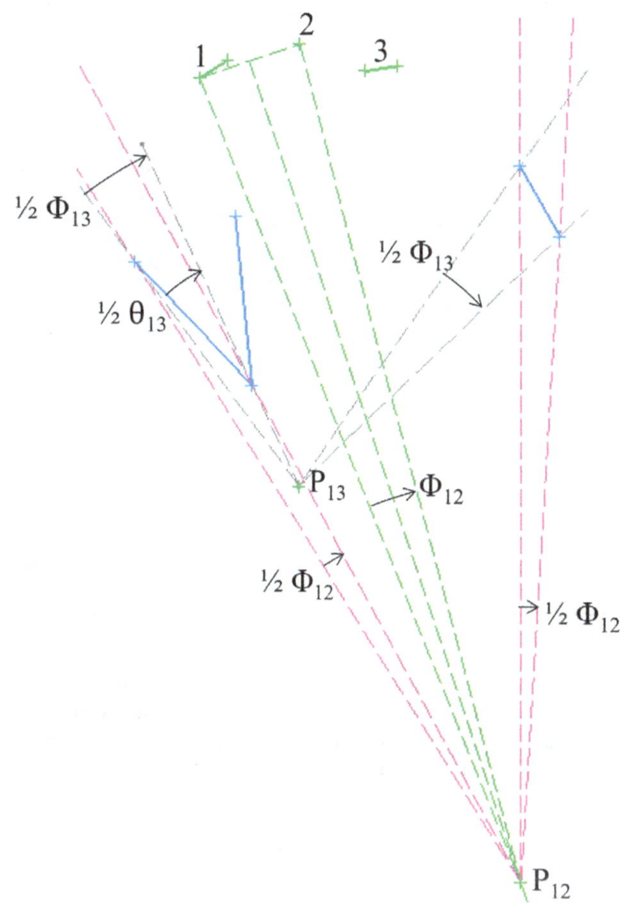

Figure 24   Solution 2-1-1 Example

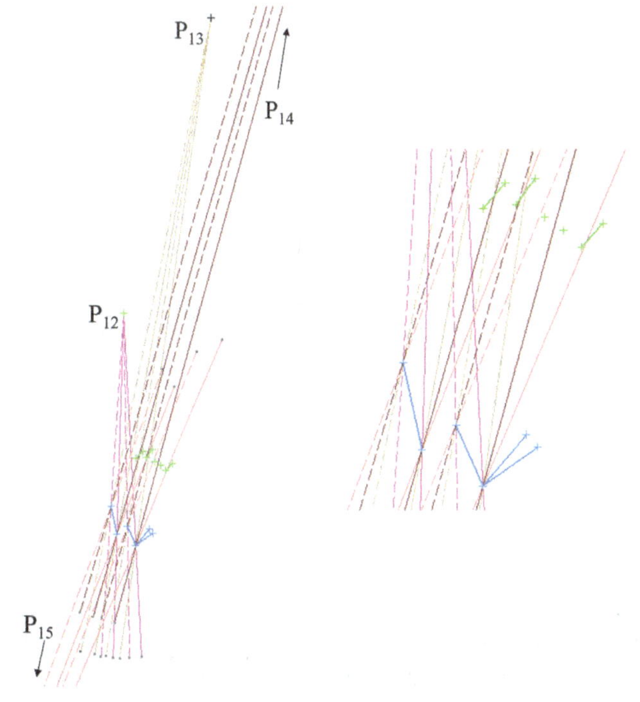

Figure 25   Mixed Synthesis Solution 3-2-2

Table 4   Mixed Synthesis Problem Limits

M	1								
P	8	7	6	5	4	3	2	1	0
F	0	1	2	3	4	5	6	7	8

M	2						
P	6	5	4	3	2	1	0
F	0	1	2	3	4	5	6

M	3				
P	4	3	2	1	0
F	0	1	2	3	4

M	4		
P	2	1	0
F	0	1	2

M	5
P	0
F	0

All fully constrained combinations are given in Table 4. Note that the corner entries in Table 4 correspond to the rigid body guidance, path and function generator problems already solved which are simply special cases of the mixed synthesis problem.

5-0-0 is the 5 position rigid body guidance problem,

1-8-0 is the 9 point path problem and

1-0-8 is the 5 position function generator problem (4 angular changes for each guiding link).

Combinations for which

$$10 > 2M + P + F$$

are underconstrained and can also be solved.

SUMMARY

Pole and Rotation Angle Constraints (PRC)

The procedure for finding solutions to any exact planar linkage synthesis problem that is not overconstrained is to ask two questions and apply one rule.

Question 1: Where are the poles?

Question 2: What are the coupler rotation angles?

Rule: The fixed and moving pivots are on lines through the pole separated by half the rotation angle.

Remember the poles must have a common subscript which is the position of the linkage on which you are doing the synthesis. You must use the moving pivot lines for that position and the rotation from the moving pivot line to the fixed pivot line is in the same direction as the rotation angle.

REFERENCES

[1] Norton R. L., 2012, Design of Machinery, 6ed. McGraw Hill, pp 99-100.

[2] Norton R. L., 2012, Design of Machinery, 6ed. McGraw Hill, pp 100-101.

[3] Hall, A.S. Jr., 1986, Kinematics and Linkage Design, Waveland Press, Prospect Heights, Illinois, pp. 8-9.

[4] Zimmerman, R. A., 2013, Planar Linkage Synthesis for Rigid Body Guidance Using Poles and Rotation Angles, Proceedings of the 2013 ASME International Design Engineering Technical Conferences & Computers and Information in Engineering Conference, Portland, OR, DETC2013-12036.

[5] Hall, A.S. Jr., 1986, Kinematics and Linkage Design, Waveland Press, Prospect Heights, Illinois, p. 129.

[6] Burmester L., 1888, Lehrbuch der Kinematik, A. Felix, Leipzig.

[7] Zimmerman, R. A., 2014, Planar Linkage Synthesis for Coupler Point Path Guidance Using Poles and Rotation Angles, Proceedings of the 2014 ASME International Design Engineering Technical Conferences & Computers and Information in Engineering Conference, Buffalo, NY, DETC2014-34058.

[8] Zimmerman, R. A., 2015, Planar Linkage Synthesis for Function Generation Using Poles and Rotation Angles, Proceedings of the 2015 ASME International Design Engineering Technical Conferences & Computers and Information in Engineering Conference, Boston, MA, DETC2015-46240.

[9] Zimmerman, R. A., (February 5, 2018). Planar Linkage Synthesis for Mixed Motion, Path, and Function Generation Using Poles and Rotation Angles. ASME. J. Mechanisms Robotics. April 2018; 10(2): 025004.

[10] Brake, D., Hauenstein, J., Murray, A., Myska, D., Wampler, C., 2016, The Complete Solution of Alt-Burmester Synthesis Problems for Four-Bar Linkages, Journal of Mechanisms and Robotics, Volume 8, August 2016.

BIBLIOGRAPHY

These textbooks provide a traditional treatment of planar linkage synthesis.

- Norton R. L., 2012, Design of Machinery, 6ed. McGraw Hill, ch. 3.

- Erdman, A., Sandor, G., 1991, Mechanism Design Analysis and Synthesis Volume 1, Prentice Hall, Englewood Cliffs, NJ, ch. 8.

- Sandor, G., Erdman, A., 1984, Mechanism Design Analysis and Synthesis Volume 2, Prentice Hall, Englewood Cliffs, NJ. ch 2, & 3.

- Uicker. J., Pennock, G., Shigley, J., 2011, Theory of Machines and Mechanisms, Oxford University Press, New York, NY, ch. 10.

- Hall, A.S. Jr., 1986, Kinematics and Linkage Design, Waveland Press, Prospect Heights, Illinois.